White-lipped pit viper

Trimeresurus albolabris

AF514078

This book sprung from my genuine interest in snakes and my curiosity about history.

Many thanks to everyone who helped me get the book from an idea to a finished product.

Thank you all for letting me use your photos! Also, thanks to those of you who uploaded images, free to use, online! All photographers are mentioned under their photos.

Thanks to Scott Eipper for excellent proofreading!

Without an understanding family, this little book would certainly not have come about, thank you Jenny, Ludwig and Jacob! I spend far too much time on herpetology, but you let me.

Finally, a big thanks to those of you who choose to read my book!
I hope you find the topic as interesting as I do!

Best regards

Rickard

Rickard Ljunggren

White-lipped pit viper
Trimeresurus albolabris

Imprint

Photo: Cover by Jonathan Hagström, photographers other pictures mentioned under each picture in the book.

Proofreading: Scott Eipper

Publisher: BoD – Books on Demand, Stockholm, Sverige
Print: BoD – Books on Demand, Norderstedt, Tyskland

ISBN: 978-91-8057-602-4

Index

1. Preamble

The white-lipped pit viper, *Trimeresurus albolabris*, is to me an iconic snake. I have a few specimens of the species at home in the basement and it is a very nice species to keep in terrariums. I have also had the privilege of having *albolabris* on my snake hook when I was in Thailand where the species is quite common in the wild.
It is, to me, simply the "original version" of a pit viper and it is often the first species that comes to mind when someone mentions pit vipers. For many in the terrarium hobby, it is also the first venomous snake they acquire. They are small, readily accessible and affordable. Since they are quite common in the hobby, there is also help and knowledge to be gained from others, especially if you choose to join a herpetological association. You can often find members in these associations who share your specific interest in the species of your interest and some members have many years of experience.
Trimeresurus albolabris is probably also the species that the tourist in Asia thinks of when they talk about pit vipers.

This book is an English version of a book I wrote in Swedish in the summer of 2023. Some additions have been made and I believe the English version is a more complete book.

The final touch of the Swedish version was written on holiday in Thailand and the following days at home. This summer will also be spent in Thailand, hopefully with a pit viper on my snake hook now and then.
It is my first attempt to write in English and I hope you like it!

Photo: Rickard Ljunggren. A male *Trimeresurus albolabris* showing off his defensive behavior. A few seconds after the picture was taken, the snake made several strikes and really wanted to bite me. If you keep pit vipers at home, you must be careful and keep your fingers at a safe distance.

Unfortunately, it is always a bit confusing with names of species when we do not use scientific names. In English, the *Trimeresurus albolabris* is often called white-lipped pit viper, but also white-lipped tree viper and sometimes bamboo viper. The name bamboo viper is shared with other species such as *Trimeresurus stejnegeri* and *Craspedocephalus (*former *Trimeresurus) gramineus*. A bit confusing and therefore I will use the name *albolabris* in this book, it is also the name I prefer to use on a daily basis.
The genus *Trimeresurus* probably contains somewhere between 35 and 50 known species, all of which are found in Asia. You can find these vipers from Pakistan, India, Japan and China to the countries of Southeast Asia, as well as on some island groups in the Pacific Ocean. According to Wikipedia, the number of species is 35 if you read the Swedish version, 44 if you choose the English instead, the French version finds 50 species and likewise do the Spanish, while in China they have settled for 43 species. In any case, these were numbers given in November 2022 when I wrote the introduction to this little book, the Swedish version. I am certain these numbers have changed after that, since then I have read about new species a couple of times*. The number of species depends on who and when you ask, and the number is in constant change it seems. Some species are divided and become several species, some species are moved to other genera and again some species are moved back to the genus *Trimeresurus* and so on.
I sometimes wonder why *albolabris* hasn't gone through the split that many species go through. It is one of the pit vipers with the largest distribution within Asia and I think it will be split into several species at some point.
*I am going through the proofreading of this text in July 2024 and right now there are 49 species in the genus *Trimeresurus* according to The Reptile Database, but *cyanolabris, kraensis, lanna* and *tenasserimensis* are not among them.

What can be said with certainty is that the number of species in this genus will change. Species will split and it is likely that we will see the genus *Trimeresurus* split again.

There is access to many beautiful pit vipers in the hobby, of which several species that were previously very difficult to obtain and difficult to afford are now both available and affordable in price. In fairness, there are species of Asian pit vipers that I think are more beautiful than *albolabris*, but a decent collection of Asian pit vipers should include at least an albolabris or two. A favorite of mine is *Trimeresurus macrops* (has changed genus but is back to *Trimeresurus* again) as I have always liked snakes with big eyes. The species' English popular name is large-eyed pit viper, which describes this snake quite well. I also have the privilege of having *Trimeresurus venustus* and *Trimeresurus insularis,* both of which are fantastically beautiful species. Enough about these and enough about me because this thin book will be about *Trimeresurus albolabris,* the pit viper of pit vipers.

Photo: Rickard Ljunggren. The female is the larger of the two, the male has his white stripe down the side of his body.

2. Scientific classification

Domain	Eukaryotes
	Eukaryota
Kingdom	Animals
	Animalia
Phylum	Chordate
	Chordata
Subphylum	Vertebrates
	Vertebrata
Class	Reptiles
	Reptilia
Order	Squamates
	Squamata
Suborder	Snakes
	Serpentes
Family	Vipers
	Viperidae
Genus	Trimeresurus
Species	Trimeresurus albolabris

Synonym
Trimesurus albolabris GRAY 1842: 48 (unjustified emendation)
Trimesurus albolabris — GRAY 1849
Lachesis gramineus albolabris — MELL 1922
Trimeresurus albolabris — POPE & POPE 1933
Trimeresurus albolabris albolabris — KRAMER 1977
Coluber gramineus — RAFFLES 1822 (nec Coluber gramineus SHAW 1802) (partim)
Trigonocephalus viridis — MÜLLER & SCHLEGEL 1842 (nec Vipera viridis DAUDIN 1803) (partim)
Trigonocephalus gramineus — CANTOR 1847 (nec Coluber gramineus SHAW 1802)
Trimeresurus gramineus — BOULENGER 1890 (nec Coluber gramineus SHAW 1802)
Lachesis gramineus — BOULENGER 1896 (nec Coluber gramineus SHAW 1802)
Lachesis grammineus (sic) — BRONGERSMA 1929
Bothrops erythrurus — LIDTH DE JEUDE 1890 (nec Trigonocephalus erythrurus CANTOR 1839)
Trimeresurus purpureomaculatus var. bicolor) — BOULENGER 1890 (nec Trimeresurus bicolor GRAY 1853, a synonym of Trigonocephalus erythrurus CANTOR 1839
Trimeresurus gramineus albolabris — MELL 1929
Trimeresurus albolabris — POPE & POPE 1933
Trimeresurus albolabris albolabris — WELCH 1994: 113
Trimeresurus albolabris — MANTHEY & GROSSMANN 1997: 407
Trimeresurus albolabris albolabris — COX et al. 1998: 20
Trimeresurus albolabris — MCDIARMID, CAMPBELL & TOURÉ 1999: 329
Cryptelytrops albolabris — MALHOTRA & THORPE 2004
Trimeresurus (Trimeresurus) albolabris — DAVID et al. 2011
Trimeresurus albolabris — WALLACH et al. 2014: 740
Cryptelytrops albolabris — CHAN-ARD et al. 2015: 284
Trimeresurus albolabris — WANG et al. 2022

Scientific name

Trimeresurus albolabris

Author Gray, 1842

Trimeresurus is Latin and refers to the "three-part tail", probably named after the tail's three-part pattern with a red line and green sides. You can see the red tail on page 50 and 63. *Albo labris* is also Latin and means "white lips", the name refers to the males' white line along the body that runs all the way to the front of the face and creates a lighter upper lip. The females lack this line and thus also the white lip as you can see on page 12.

Author names are, within biological systematics, the name of the person or group of persons who first used a particular scientific name for a particular taxon, such as a species or other biological entity. This must also be done in accordance within the rules outlined by the ICZN. Author goes back to the Latin *auctor* which means, among other things, "author" and "originator".

For this books *albolabris*, the author is John Edward Gray (12 February 1800 – 7 March 1875) who was a British zoologist. He was the older brother of the zoologist George Robert Gray and the son of the pharmacologist and botanist Samuel Frederick Gray. Gray was born in Walsall, which is roughly between Birmingham and Wolverhampton.

Photo: Wikimedia commons, photographer unknown.

The motif is an engraving made by Thomas Herbert Maguire in 1851 and depicting J E Gray.

While growing up, Gray's family moved to London, where he came to study medicine. He began his zoological career by volunteering to collect insects for the British Museum at the age of 15.

After being "blackballed" (voted out) by the Linnean Society of London, Gray shifted his interest from botany to zoology. Balloting (French *ballotte*, meaning little ball) is a voting procedure with balls and is used in closed polls. Traditionally, black and white balls are used, with white indicating a yes vote and black a no vote. In several orders, guilds and more, a balloting that participates in only white marbles, that is, a hundred percent positive outcome, is called "brilliant". Perhaps not brilliant of me to include this extremely irrelevant fact in a book about pit vipers but at least now you have learned something new from reading my book about a specific Asian viper.

Gray officially joined the Zoological Department in 1824 to help John George Children catalog the reptile collection. In 1826 he married Maria Emma Smith who helped him with his scientific work, especially with drawings.

In 1840 he took over the position of Keeper of Zoology at the British Museum in London. He held the post for almost 35 years and he published several catalogs of the museum's collections. These catalogs contained extensive discussions of animal groups and descriptions of new species, and he improved the zoological collections so that they became among the best in the world. During his 50 years as an employee of the British Museum, Gray wrote almost 500 papers, including many descriptions of species new to science. These species had been presented to the museum by collectors from all over the world and included all branches of zoology.

He described more than 300 species and subspecies of reptiles, surpassed only by his successors at the British Museum, George A. Boulenger and Albert Günther, and the American zoologist Edward D. Cope.

John Edward Gray is buried in St Mary's Church, Lewisham.

The genera named in his honor include:
The snake genus *Grayia* with four known species.

Species and subspecies named after him include:
The gecko *Naultinus grayii.*
The skink *Ablepharus grayanus.*
The skink *Tropidophorus grayi.*
The turtle *Trachemys venusta grayi.*
The lizard *Delma grayii.*
The lizard *Microlophus grayii.*

In this text, I have only included examples of species and genera from the world of reptiles. It is also not impossible that I missed some species from the world of herpetology, so to be safe, we must see the list as examples of species named in his honor and not as a complete list of species with his surname in the species name.

The genus *Trimeresurus*

The genus *Trimeresurus* was first described by the French naturalist Bernard Germain de Lacépède in 1804. He was an early evolutionary thinker and argued for constant change of species. He believed that species change over time and can die out from geological catastrophes or undergo metamorphosis into new species.

Most of the species in the genus are green and many species are quite difficult to distinguish morphologically.

Photo: Rickard Ljunggren. A male with his bright stripe down the side of his body. It's rather easy to tell when an *albolabris* wants food. They will then lie on a branch, head down, watching the ground below

Most of the species in the genus are quite small vipers, about the size of a *Vipera berus* or a little bigger, but a *Trimeresurus sumatranus* can grow up to 160 cm. The various species lives mostly in trees, but there are also some ground-dwelling species. Many species have the typical red tail, but not all of them. Just like many vipers and colubrids, they can sometimes vibrate their tail as a warning when they feel threatened. Most species in the genus are ovoviviparous. The reproductive biology of some species is still unknown.

The different species in the genus eat a variety of animals, the most common being various lizards, amphibians, small birds, rodents and other small mammals. In captivity it is common to give pit vipers mice, but sometimes it is preferable to get juveniles started by offering them small geckos instead.

The genus Trimeresurus is found in Asia from the Indian subcontinent (including Sri Lanka), throughout Southeast Asia, China, Japan and the Pacific Islands. The distribution area of the genus may change when species are added or moved from the genus.

All pit vipers are venomous, and some species can be quite dangerous for us humans. Death after a bite from one of the species in the genus is, however, rather unusual. In Thailand and parts of Indonesia, they are responsible for a large proportion of the countries venomous snakebites. Partly because many *Trimeresurus* species are fairly common in their habitat, and partly because they are fairly small snakes that are quite often found in or near urban environments.

The species is undoubtedly the most common venomous snake in Thailand's capital along with *Trimeresurus macrops*. These two species can sometimes be found side by side, and for many people it is hard to separate them from each over. In Thailand, the two species account for approximately 40% of all venomous snake bites that lead to hospitalization. In Bangkok, the two pit viper species account for 95%! You can come across these pit vipers in Bangkok's leafy parks, gardens, small bushes, abandoned lots, trashy alleys, railway area, small garbage dumps and along waterways. Actually, anywhere where there is enough nature to offer protection and at the same time the opportunity to take prey sometimes.

The venom of *Trimeresurus* species varies in toxicity, but all are primarily hemotoxic (but also cytotoxic) and are considered medically significant to humans.
Hemotoxic venom has effects on the heart and blood vessels.
Cytotoxic venom has local cell-killing effects around the bite.
Just like other vipers, pit vipers deliver painful bites and according to those who have been bitten, the venom is really not a pleasant experience. A bite can lead to necrosis around the bite area and it takes a long time to heal. Always seek medical attention if you are bitten. Example of a bite from a *Trimeresurus albolabris* can be found on next page.

During the Vietnam War, American foot soldiers had an exaggerated view of the toxicity of pit vipers. The little green snakes were feared and fabled. You can read more about this in the next chapter.

The severity of a bite is often impossible to predict as it depends on many factors such as the species and size of the snake involved, how much venom was injected (if any) and the size and state of health of the person bitten before being bitten. The bitten may also be allergic to the snake venom or the antidote.

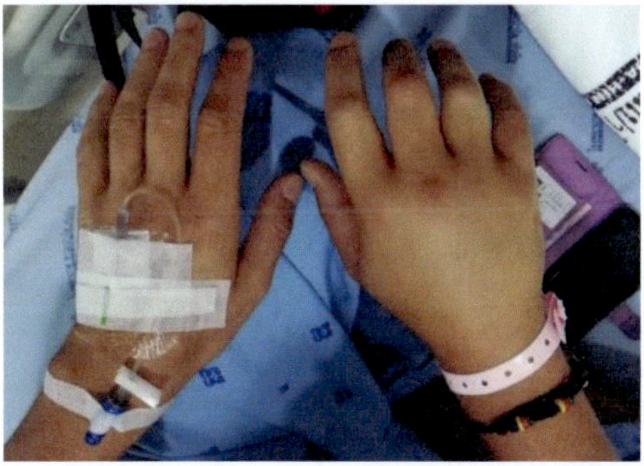

Right hand bitten by a *Trimeresurus albolabris.* The visual symptoms can be seen as typical of a bite from a viper species. Jonathan Hagström, who was bitten, sought medical attention immediately. The hospital staff initially decided not to give him an antidote because it is usually not needed for this particular species of snake. Three hours later a new decision was made, and Jonathan received a total of five doses of serum. In addition to serum (which he turned out to be allergic to), Jonathan received adrenaline (given for allergic/anaphylactic reactions or circulatory shock), antibiotics (given only for symptoms of infection) and morphine (it hurts!)

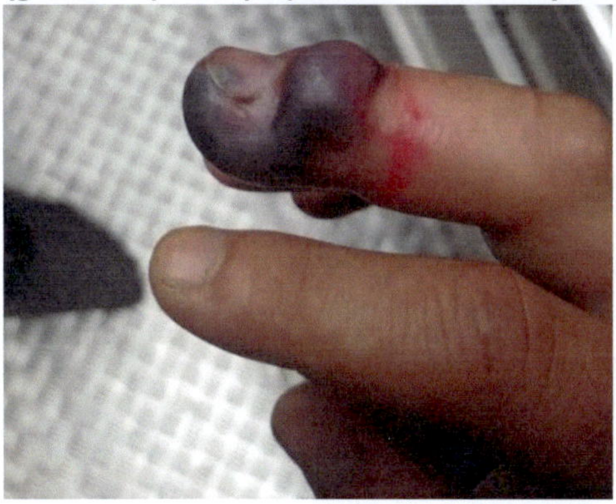

during his hospital stay. The finger was restored after about two months. Photos: Jonathan Hagström with friends. Upper picture the same day as the bite, lower picture three days after the bite.

The family *Viperidae*

Viperidae (vipers) is a family of snakes found in most parts of the world, except for Antarctica, Australia, Hawaii, Madagascar, New Zealand, Ireland and some other islands that were isolated before the time of the vipers. In some parts of the area of distribution, vipers are the only naturally occurring venomous snakes, as for example in Great Britain and here in Sweden where I live.
Vipers are currently found in almost every available habitat on Earth. They are found in both temperate and tropical environments and in a variety of habitats. Vipers are found in environments where life is lived exclusively on the ground as well as environments where life is largely lived in bushes and the branches of trees. Vipers live in all environments from jungle, rainforest and forest to grasslands, dry steppes, rocky mountain environment, savannah and desert-like sandy environment. Some species have also adapted and can live close to us humans, even in metropolitan environments.

The oldest known fossils of vipers are from the Miocene epoch, about 23 to 5 million years ago, but research suggests an older origin for the Viperidae dating back to the early Eocene epoch, which spans from around 56 to 34 million years ago. The vipers first spread in the Old World (Europe, Africa, Asia) more than 50 million years ago and then continued to spread further into the New World (America and Oceania). Several interesting features emerged during their evolutionary history, including the heat-sensitive pits, the care of offspring found in some species, and specialized sound production in rattlesnakes. They developed the ability to eat unusually heavy and bulky prey and gradually acquired an impressive array of dietary specializations. Although vipers represent an interesting and successful distribution, many species are threatened with extinction and several aspects of their evolutionary history still needs to be investigated.

The family *Viperidae* has three subfamilies, with a total of just over 200 different species of venomous snakes. Some sources suggest a fourth subfamily, *Causinae,* but many consider the snakes of the genus *Causus* (African night-adders, currently seven different species) to belong in the subfamily *Viperinae.*

The subfamily *Viperinae* can be found in Europe, Asia and Africa. The European vipers belong to this subfamily, but also slightly larger and more potent snakes, for example *Daboia russelii* (Russell's viper) *and Bitis gabonica* (East African gaboon viper).
Crotalinae is probably the most well-known subfamily. These snakes can be found from Eastern Europe east through Asia to Japan, Taiwan, Indonesia, the Indian Peninsula and in Sri Lanka. You will also find these snakes on the other side of the Atlantic, from southern Canada south through Mexico and Central America to southern South America. Here you will find the main character of this book, the *Trimeresurus albolabris.*
Azemiopinae is the third and smallest subfamily. Species within this family are found in Myanmar, southeastern Tibet, southern China and in northern Vietnam. In this subfamily you will only find the genus *Azemiops* which contains the two species *Azemiops feae and Azemiops kharini.*

The name *Vipera/Viperae* is Latin and means "viper". The name possibly has an origin in the Latin words *vivus* (living) and *parere* (to give birth). The name may refer to the fact that vipers give birth to live offspring and do not lay eggs. Exceptions exist, there are egg-laying snakes that nowadays belong to the viper family. Most species are ovoviviparous, which means that these animals possess embryos that develop inside eggs that remain in the mother's body until they are ready to hatch. The number of juveniles in a litter usually remains constant through the mother's life, but as the female grows larger, the eggs grow larger and contains offspring that are larger at birth.

General description of vipers.

Some vipers' patterns help the snake to facilitate escape, the pattern confuses and makes it difficult for the threat to determine the snake's speed and where it is going. If you have seen a *Vipera berus* while mooving, you have probably seen the phenomenon.
The vipers' tail can be very short, as in rattlesnakes, or relatively long as in arboreal viper species. Arboreal vipers also have a prehensile function in the tail, adapted for their arboreal lifestyle. Members of the viper genera *Crotalus* and *Sistrurus* are known for the rattle at the end of the tail, which consists of a series of hollow skin segments. This rattle can be vibrated or shaken to make a rattling or buzzing sound. Many other viper species also vibrate their tail when they feel threatened but need the help of the substrate to create sound. Right now, I have five specimens of *Trimeresurus albolabris,* all of them vibrate their tail when they want to inform me that my presence is unwelcome.
Most vipers have keeled scales (the scales are not smooth, they have a ridge in the middle that often looks like a line in the scale), a fairly stocky build body, a rather short tail compared to many colubrids, and a triangular head distinct from the narrow neck. On page 54 you can see a picture of keeled scales.

The eye of a viper.

Vertical pupils aren't just there to improve night vision in vipers and other snakes. Research shows that the narrow pupils can also help the snakes when ambushing or stalking prey as the snakes depend on not being detected because the prey animals in many cases have an escape speed that the snakes cannot keep up with. The same research shows that snakes that lie in ambush or hunt at night tend to have vertical pupils, while round pupils are more characteristic of diurnal snakes that actively seek and stalk their prey. But overall, the narrow shape of the pupils was more strongly associated with hunting behavior than with nocturnal activity. By studying the evolutionary relationship between the different species, it was found that vertical and circular pupils evolved several times in different groups of snakes and that hunting behavior was the strongest driver of pupil shape.

The narrow, vertical pupil probably helps ambush-hunting snakes pursue prey at night by making objects at some distance appear sharper from the snake's hiding place. Just like a smaller aperture on a camera lens, a smaller pupil creates a deeper field of focus. Vertical pupils are only smaller in the horizontal plane, meaning they offer a greater depth of field in the horizontal plane but still allow enough light into the eye for night vision. Having a focused field of vision can help an ambush predator as it reduces the need to move forward towards the prey to focus on the prospective meal – thus reducing the risk of detection.

There are also theories that the vertical pupil aids in the snake's camouflage by mimicking surrounding grass and vegetation. The vertical pupil may also help reduce the risk of glare, which may explain why the pupils appear rounder in light.

Photo: Rickard Ljunggren. A female of the species *Trimeresurus macrops* that missed when striking after the food I offered and bit the terrarium's vegetation instead. Note the pupil which is narrow like a line.

The venom and the teeth of a viper.

All species in the family are venomous and have two fangs in the front of the mouth which are very long in relation to the size of the snake. The size of a snake species' fangs is not proportional to the danger of the species' venom. For example, the South American pit vipers in the genus *Lachesis* have enormous fangs, while much smaller fangs are found among the potent snakes in the genus *Pseudonaja (*family *Elapidae)*. The species in the genus *Lachesis* are very venomous but drop by drop the venom of brown snakes (*Pseudonaja)* is considerably more potent.
Many vipers' venom is primarily haemotoxic. Haemotoxic venom takes more time than neurotoxic venom to immobilize prey, so vipers must track the prey they just bitten through a process called prey relocation. The exception is the tree-dwelling vipers that hold onto their prey after administering a venomous bite.

> Basic venom types:
>
> Haemotoxic (= haemorrhagic) venom mainly have effects on the blood vessels.
>
> Neurotoxic venom have effects on the nerve system and the brain.
>
> Cytotoxic venom have local cell-killing effects at the site of the bite

Vipers can track their bitten prey via certain proteins found in their venom. This important adaptation allowed rattlesnakes to develop the hit-and-release bite mechanism, which provided a huge advantage by minimizing contact with potentially dangerous prey. The venomous snake's only weapon is its fangs, and if these are stuck in a prey, it is not easy to defend itself against any prey with beaks, teeth or claws.
This adaptation thus requires the snake to track down the bitten animal, in order to eat it, in an environment full of other animals of the same species and many other species.

Tests have been done on rattlesnakes and it was found that they more easily located mice injected with the venom of the species in question.

The venom also helps with digestion, and you can therefore say that digestion kicks into gear as soon as the venom is injected. This is an important feature because many vipers have inefficient digestive systems.
The most important factor regarding the amount of venom injected is the size of the snake, larger specimens can deliver a much larger amount of venom. The species is also important in terms of amount of venom as some species are likely to inject more venom than others, may have more venom available or may deliver more bites in a single strike.
Viper venom usually contains an abundance of protein-degrading enzymes that produce symptoms such as pain, severe local swelling and necrosis, blood loss from cardiovascular damage, and disruption of the blood coagulation system. Some viper bites produce neurotoxic symptoms.
Death is usually caused by blood pressure becoming so low that it leads to organ failure.
Bites from vipers are often a very painful experience and should always be taken seriously, even when not necessarily fatal. Even with prompt and proper treatment, a bite can still result in a permanent scar, and in a worst-case scenario, the affected limb may even need to be amputated. Even after a snake is thought to be dead, fangs can still inject venom. Snakes believed to be dead have bitten many people and have caused serious, or even fatal, envenomations.

Vipers are among the most famous and probably most medically important venomous snakes. In addition, they are fantastically beautiful and interesting snakes! However, one must never forget that a mistake in the vicinity of a viper can have serious consequences, even though many people who gets bitten make a full recovery.

Viperid envenomation results in tissue loss, deformity or loss of function in a significant proportion of those bitten. The proportion naturally varies from species to species.

Also remember that a viper's strike can be incredibly fast, and they can reach farther than you might think!

Photo: Rickard Ljunggren. A male *Trimeresurus albolabris* who made a couple of defensive strikes. On bottom right you can se my warning signs that I will take with me if I need to seek medical care after a bite. "Varning giftorm" in Swedish means "Warning Venomous snake". The red area on the map is the distribution area of *Trimeresurus albolabris.*

All vipers are solenoglyph*, with long fangs on the front of the upper jaw that resemble hypodermic needles. The ability to fold their fangs back towards the palate means that vipers can have longer fangs than other snakes. Snakes such as cobras have fangs that mainly are in a fixed position with a restricted degree of rotation. With the help of individual muscles, most venomous snakes can articulate their fangs together or individually. During a viper strike, the mouth can be opened to almost 180°, the upper jaw rotates forward, and the snake rotates the fangs forward as late as possible so that they are not damaged (as they are rather fragile). The long fangs of vipers means that the fangs penetrate rather deeply.

* Having solenoglyphic teeth means that the fangs that transport the venom into the prey are hollow. The fang simply looks like a needle.

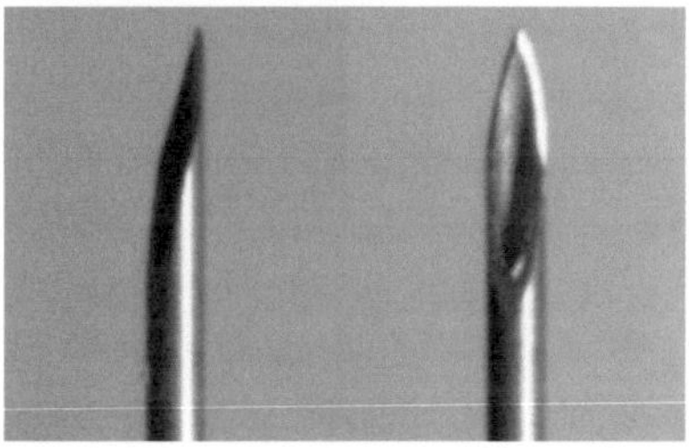

Photo: Fifo, Wikimedia Commons.
Image of cannula in two different angles. A viper's venom fang looks the same. (Image cropped.)

The jaws of vipers close when in contact with the prey/threat and the muscular sheaths encapsulating the venom glands contract and pushing the venom down the venom duct and through the fangs as they penetrate the target. This strike is extremely fast and in the case of defensive strikes more than one bite can occur. In defensive attacks, the vipers sometimes deliver dry bites to save venom. In addition, to having the ability to deliver dry bites in defense, the vipers can also dispense venom when biting their prey. Larger quantity for larger prey and smaller quantity for smaller prey. It is even the case that vipers can regulate the amount of venom according to the species of prey or predator (threat)! Nature is fantastic!

3. White-lipped pit viper as a weapon?

"Any veterans out there who encountered snakes during military service?"
That was the beginning of a thread on Facebook in a group whose topic is the Vietnam War that lasted from 1955 to 1975. I found the thread a few years before the progress of writing this book was started in 2022 (the Swedish version) and there somewhere the idea of a book about the *albolabris* was born.

Photo: Pixabay. American helicopter evacuates soldiers in Vietnam.

Those who read the thread got a thrilling insight into what it could be like for an American soldier far away from home, in the deep jungles of Southeast Asia. The post about snakes during the war also included a picture of some soldiers holding up a large snake that was obviously dead. A war veteran who responded to the thread identified the snake as a Burmese python (*Python bivittatus).* Just like in so many other contexts, the size of the snake was exaggerated. The snake was said to be six meters long but looked a bit smaller to me.

It is a well-known fact that the Viet Cong used various types of traps as part of the warfare against American soldiers during the Vietnam War. The Viet Cong was an epithet and umbrella term to call the communist-driven armed movement and united front organization in South Vietnam. They fought under the direction of North Vietnam against the South Vietnamese and United States governments during the Vietnam War.

However, the use of snakes in these traps is a controversial topic. There are reports suggesting that the Viet Cong used pit vipers in their traps, but there are also theories that pit vipers in the traps originated solely as propaganda to scare American soldiers.
There are some stories about soldiers encountering snakes in traps during the war, but it is unclear how often this actually occurred. Furthermore, it is difficult to determine whether the Viet Cong deliberately placed snakes in the traps or whether the snakes were simply a natural occurrence in the dense jungle environment in which a rather big part of the war was fought.
In conclusion, there are some reports suggesting that the Viet Cong may have used snakes in their traps during the Vietnam War, but it is unclear how widespread this practice was or how effective it was as a military tactic.

The reason the Viet Cong used traps was to slow the advance of the American soldiers in the dense jungle terrain in a cheap way. These traps were often very simple and made from readily available materials, such as sharpened bamboo poles or glass bottles, which the Viet Cong buried in the ground or hung from tree branches. When triggered, they would impale the American soldiers and cause serious injury or even death. The Viet Cong also used trip wires that set off a grenade or some other explosive device, and they sometimes mined the corpses of American soldiers. The US army had an economic advantage over the Viet Cong, this was a cost-effective warfare for the Viet Cong.

The fear of traps was quite common among American soldiers and meant that they moved forward with more caution than they would have done without the risk of stepping wrong somewhere in the jungle.
What, then, were the mentioned traps that may fall within the scope of this book? Rumors circulated that the Viet Cong sometimes hung pit vipers in their tunnels in the jungle, and stories circulated of tunnels with rigged traps. When the traps were triggered, a pit viper was dropped from the ceiling, landing on the back or head of the American soldier who happened to trigger the trap. In some tunnels there would be pit vipers suspended at face height that you risked not seeing down there in the dark. Several people in the thread I read talked about these traps and suspended pit vipers, but no one talked about their own experiences. The question is if anyone really had this happen to them or if it is a myth that has been given life in folklore?

" I worked with a guy who was a tunnel rat in Vietnam. He told me about snakes being tied to the walls of a tunnel, but usually the snakes died before the troops ever got there."

"I am a Vietnam vet and the 2-step snakes were by far the most scary to me. I think that is because of my innate fear of snakes was already present when I went in-country. Although, the bullet trap is second."

"Yes, this is exactly what it sounds like. Viet Cong guerrillas would often carried Bamboo Pit Vipers in their packs to (hopefully) kill anyone who searches through them. They would also tie the deadly snakes to bamboo and hide them throughout their tunnel complexes. When the Bamboo was released, so was the snake – right onto the enemy. "

" Charlie used to tie them by the tail, put them on a shelf above the doorway openings in tunnels. First thing that moved under

that opening got hit! In the head! Never get me to be a tunnel rat."

Four different veterans who tell their story, but do not talk about their own experiences.

Photo: Wikimedia commons, Signal Corps photographer.

In this picture you can see the tunnel rat Sgt. Ronald H. Payne with a flashlight and a gun. The picture is taken in Vietnam in 1967.

The tunnels in the jungle were originally dug as hiding places during the freedom struggle against the colonial power France, in what is known as the first Indochina War (1946–1954). During the Vietnam War, the tunnels were repaired, and new ones were dug. The 48 kilometers of tunnels that existed from the time of the French colonial war became 200 kilometers of tunnels during the war against USA. The tunnels were used by the Viet Cong mainly as hiding places in connection with battles, weapon caches, accommodation for the combatants, infirmaries and as communication and supply routes. Today it is a popular destination for tourists. The tunnels were adapted for the Vietnamese and thus cramped for most of the American soldiers. The tunnels that are visited by tourists today are often widened so that we Westerners can take a tour, but when a Swedish friend of mine visited them there was a rather big American tourist stuck in a tunnel.

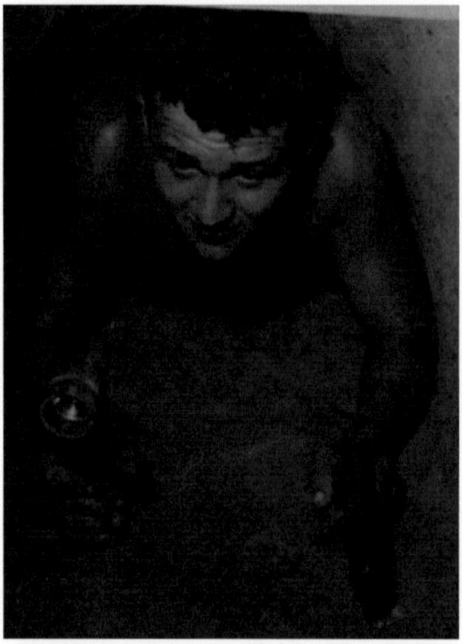

Photo: Wikimedia commons, Signal Corps photographer.
In the picture you can see the tunnel rat Sgt. Ronald H. Payne with a flashlight and a gun. Picture taken in the tunnels in the Vietnamese jungle in 1967. The American soldiers whose job it was to search the Vietnamese tunnel system and look for "Viet Cong" were often called tunnel rats.

The tunnels were often used by Vietnamese soldiers as a way to appear out of nowhere, where entrances and exits were hidden, and the Vietnamese could surprise soldiers who were unknowingly staying above or near a tunnel. See picture on the previous page, of a found tunnel with an exposed entrance.

It sounds like a very thrilling story. Deadly traps with insanely venomous snakes, deep into the unknown Asian jungle, under the ground in narrow tunnels filled with enemies. But it's probably not much more than a rumor. I find evidence of soldiers being bitten by snakes, but nothing to support that an enemy staged some kind of trap for it to happen. But maybe it happened somewhere, sometime? Not everything that happens in a war is documented down to the smallest detail.

"Asked our flight surgeon if there was any truth to this 2 step bamboo viper thing. He said let me put it this way. If one bit you right now, all I could do is watch you die."

Stories of viper-filled traps the soldiers could fall into also occurred but there is no confirmed record of anyone in the Viet Cong building a single snake pit during the war. -They probably had better things to do. The amount of time and effort it would take to possibly get a couple of soldiers bitten by a snake makes the idea fall on its own absurdity. But the myth of snake pits in the jungle has certainly been an obstacle for some American soldiers.

The Facebook group has over 30,000 members, many of whom were American veterans from the war in Southeast Asia. Few threads in the group seemed to engage as many people as the post about pit vipers in Viet Cong traps. In three days, the post passed 200 comments and many of the comments were about the small light green snakes that most veterans called "two stepper," or "two step."

The veterans portrayed the pit viper, which was commonly called the "bamboo pit viper" as a very dangerous and extremely venomous snake. The story among American troops in Vietnam was that if the pit viper bit you, you could only take two steps before you were dead. Hence the pit viper got the name "two stepper" or variants of this name. Of course, a pit viper bite can be fatal, even if it is rare, but in the world of snakes, death after a venomous bite is never this quick.

In total, there are about 200 different snake species in Vietnam and about a quarter of these are more or less venomous.

In the country's fauna there are several other venomous snakes that may be considered considerably more venomous than any of the pit vipers, but it was the green pit vipers that became the most notorious. Maybe because it was harder to spot them in the green jungle and more soldiers happened to get too close? The kraits, which are extremely venomous, were called "three step", which should mean that they were considered almost as venomous as a pit viper.

Photo: Pixabay. American soldiers in the field, somewhere in Vietnam.

US archives and other sources indicate that between 25 and 50 American soldiers per year were bitten by venomous snakes during the Vietnam War. In total, roughly 8.7 million American soldiers fought in the Vietnam War. In other words, snake bites were not that common during the war.
There are no figures on how many died from snake bites. Roughly 900 soldiers died of disease during the war, but whether any of these deaths were due to snake bites is not clear from the statistics. In total, roughly 58,000 American soldiers died during the war, which is terrible.
There are no exact figures of how many Vietnamese died during the war. According to a statement by the Vietnamese government in 1995, it was estimated that 5.2 million, or one in eight inhabitants, had been killed during the war. An awful number. What could be worth over 5 million human lives? The Vietnamese figures also include a large proportion of the civilian population.
War is truly terrible, but history always repeats itself. The thirst for money and power will get someone to sooner or later starting a new war.

Although snake bites were quite rare to experience, pit vipers became a popular psychological weapon to use by the Viet Cong. Whether it was a biological weapon or just propaganda, it had some effect. It was probably not exactly the *Trimeresurus albolabris* that was so feared, but green pit viper species in general (genus *Trimeresurus*). American soldiers couldn't tell the difference between an *albolabris*, a *macrops*, or any other of the little green pit vipers, and neither could most of the Vietnamese soldiers.

Of course, what is true and what is a legend cannot be known, but the fact that there was a fear of Vietnam's venomous snakes among many American soldiers who fought in a foreign jungle can probably be considered true. Presumably it was the fear of a venomous snake bite that was a greater obstacle to American soldiers than the actual venomous snake bites. The thought of running into a "two stepper" possibly slowed down and hindered American soldiers who had to make their way through Vietnam, actually running into one of the feared pit viper traps was probably extremely rare, if it ever happened at all. The phenomenon is somewhat talked about, both during and after the war, but most stories are retellings where someone tells what they have heard about.

To me, it is more like an urban legend than a narration of real events. My guess is that the biggest effect this biological weapon had was psychological. My conclusion is that the tale struck fear into the foot soldiers who had to make their way through the jungle and countryside in Vietnam, but no one saw any traps with pit vipers.

My guess is that, as so often in the world of snakes, the origin of the tale is found in fear and ignorance. Several veterans' stories show ignorance about snakes and their behavior. A soldier had caught a small monocle cobra, *Naja kaouthia*, and kept it as some kind of pet for a while. He was aware that a young cobra is just as venomous as an adult, but he says that the young cobras have their teeth "embedded" and had to chew in the venom. Therefore, it was possible to pat it, but not too much. A statement you should not take seriously, no cobra needs to chew in its venom. The truth is that newborn cobras bite just as effectively as adults.

Some retellings semes saner to me, like these:

"I was crew chief flying Dustoff (a medical evacuation chopper)," wrote Neal Caspersen of Pensacola, Fla. "I picked up three in a year bit by the little green vipers; all were alive when we dropped them off. One was a Vietnamese girl about 7ish.

"When we loaded her someone handed me a Styrofoam coffee cup; in it was the dead little green viper, the color was striking and it was so small."

There are some reports about American soldiers who were afraid of snakes during the Vietnam War, but this fear was not widespread or universal. Many soldiers encountered venomous snakes while serving in Vietnam and it is possible that some developed a fear of them as a result. Especially if they were bitten or witnessed others being bitten.
Fear of snakes varies from person to person and may be due to a variety of factors, such as previous experiences with snakes or personal phobias. Rumors of snakes in the jungle and traps with snakes may have fueled the fear for some. While it is possible that some soldiers may have been afraid of snakes or have developed a fear after encountering them in the jungle, fear of snakes was probably not a major concern for most soldiers fighting in an already very challenging and hazardous environment. In addition to the physical danger posed by snakes, soldiers may also have experienced psychological

distress due to the unfamiliar and hostile environment of the war zone. All in all, the fear of snakes among American soldiers during the Vietnam War was probably a relatively minor issue in the broader context of the conflict, but if you read in forums today, the myths about the deadly snakes are alive and well. Almost 50 years after the end of the war!

We'll stop there. It turned out to be a rather long chapter about something that may never have happened. But an exciting story nonetheless and the pit vipers played their role as a psychological weapon regardless. Of course, there was the risk of encountering snakes in the tunnels, as snakes like to seek safety below the surface of the ground and sometimes even cool off if it's warm. I came across a non-venomous *Elaphe taeniura ridleyi* (Ridley's beauty snake) about 100 meters into a cave located near Hua Hin, Thailand.

In some American soldiers the little green snakes created a fear which certainly may have been a hindrance or a deterrent at times, so far, we can be sure. The little green snakes seem to have been feared by some and discussions I read online while working on this chapter testify that the pit vipers have almost become mythical.

The post that I refer to at the beginning of this chapter has today been deleted. I made some attempts to start up new discussions on the topic. In some forums my attempts were deleted, not sure why. In other forums, the post was allowed to remain, but the answers I received were not entirely relevant and no one told me about their own experiences with green snakes in traps or in tunnels. The answers I got confirmed that the rumor was true but only repeated what they had heard from other people talking about their experiences or what they have heard.

The legend lives on.

4. Tourism in Southeast Aisa

Traveling is a wonderful pleasure, whether you seek relaxation or want to discover the world. I like to travel to Southeast Asia as often as my wallet allows me to. Nowadays, I am a more conscious tourist than I used to be and many people with me have started to think about how they spend their money when they are on holiday. More and more people are becoming aware that with our tourism and our Western purchasing power we have a big impact on the tourism industry and the environment, especially at popular tourist destinations.
As an animal-friendly tourist, you should make demands and influence the tourism industry's view of animals. What you say and do can have some positive effect for the animals, but how you spend the holiday fund can have a very big effect for the animals. The supply of goods and services at a destination is governed by demand and as a tourist you should use your consumer power to make a difference.

You can do that by:
*Not buying souvenirs that are made of animal parts.
*Avoid entertainment and transport that uses animals.
*Avoid arrangements where you get to hug, hold or be photographed with an animal.
*Avoid zoos or similar with substandard animal husbandry.

Sometimes it may be difficult to decide what is good or bad, especially if you do not happen to have a greater interest in animals and nature. To make it easy for you, if you're not sure if it's okay or not, choose to opt out. If it had really been okay, you probably wouldn't have doubted it. Sometimes it can certainly be difficult to refrain. It is a "once in a lifetime".

It might feel very thrilling with dangerous animals, cute and exciting with small tiger cubs or fantastic with an elephant that can paint a picture. But still choose to refrain!
Your vacation memory which you experienced for 45 minutes, or your awesome vacation photos took place at the expense of the animals. Both at the expense of the animals that were part of the entertainment and at the expense of the animals that will be part of the entertainment in the future when today's animals are worn out and used up.

We are all aware of the debate surrounding Europe's circuses and the animals that have been harmed there in the past. We are all appalled when we see advertisements for animal rights organizations on television that show clips of bears in far too small cages or animals being beaten to be broken down and useful for entertainment.
Unfortunately, some people forget this awareness when they are far from home. You can still see on social media someone happily posing with a chained and drugged tiger or someone happily riding on the back of an elephant. In the city's crowds outside the nightclubs, someone is posing with an iguana, a monkey or a python and they pay to take the exciting selfie to show off to those at home. I have contributed to this in the past but today I wouldn't dream of sponsoring this form of tourist attraction.
The gratifying thing is that it is not just me who has had an awakening, but more and more people are choosing to distance themselves from these tourist attractions. You simply no longer want to have a picture with that drugged tiger or something similar. Unfortunately, this awareness has not reached all the way to the snake shows. They may have different names, but regardless of whether it is called "Snake show", "Cobra show" or "King cobra show", it is as unethical as the dancing bear in old-time circuses. My most unscientific belief is that it's easier

to empathize with that little chained monkey or tiger cub that people line up to pet. If you google unethical animal tourism, you will find many articles about why you should not contribute to elephants, tigers, birds and monkeys being used as entertainment for tourists, but you will not find much text that addresses the reptiles' conditions in the tourism industry.
You see the same phenomenon in Sweden, my home country. Homeowners can still unashamedly chop up a snake that visits their property, sometimes you even see people taking pictures and displaying killed snakes as a trophy on social media. Those people would never think of doing the same to visiting furry or feathered animals. Of course, both parts are completely unreasonable because even if we don't think the animals are cute, we don't have the right to kill them simply because they are on our property.

Pit vipers are rarely used in snake shows. There are probably two reasons why they can avoid this. Partly because they are quite small snakes, it is simply difficult for an audience to see that little green snake. Partly because an ordinary tourist has no idea what kind of snake is being displayed. It is simply more spectacular with an iconic cobra showing its hood and everyone knows how dangerous a cobra is. The only time you see a pit viper in these contexts is when they are displayed to the public on a snake hook or a stick during some form of halftime pause in the entertainment. Someone walks around and shows the green snake to the seated audience.
Of course, these are wild-caught animals and if they don't survive long in captivity, you can always look for a new snake. Snakes are captured nearby and the animal husbandry is quite often substandard and directly life-shortening for the animals.

Photo: Rickard Ljunggren. Picture taken in Phuket, Thailand, many years ago, before I had the sense not to visit snake shows. Picture showing a monocled cobra (*Naja kaouthia*) at a cobra show. In addition to animals getting hurt, safety is often substandard for us humans. Accidents happen, even fatal ones. Most often, it is the "artist" who gets a bite.

Photo: Rickard Ljunggren. A *Trimeresurus albolabris* is being displayed at a show at Queen Saovabha Memorial Institute in Bangkok. Picture taken with a cell phone 2011, hence the quality.

Most of the species used in these shows are viable, but that doesn't matter to the individual animal who is part of the entertainment. The animals that the tourists choose to be entertained by live under high stress in an environment that usually does not take the snake's best interests into account. You should not buy souvenirs made from animals or animal parts. The main reason, in my opinion, is not that it probably is illegal to bring these souvenirs through customs. I believe that the main reason is that the animals are harvested in the wild to become an exciting ornament on someone's bookshelf.

Photo: Rickard Ljunggren. Unfortunately, I photographed through the shop window of the pharmacy, hence the poor quality.
Photo taken with a cell phone in 2011 in Guangzhou in southern China. Bottle of natural medicine, filled with snakes and lizards. Folklore is not easy to change, but our way of tourism should be easy to change.

Most often it is said to be a cobra in these bottles, but there are variations. A study from 2009 showed that about 10% of the bottles did indeed contain a cobra, but close to half contained the species Yellow-spotted keelback, *Fowlea flavipunctata.*

Photo: Pixabay.

The yellow-spotted keelback is not a cobra, it is a completely harmless snake species that often is harvested to become souvenirs. Perhaps it is not entirely unreasonable that you would rather catch harmless snakes than shy and highly venomous cobras when you are making souvenirs.
The snake's body is stretched out with steel wire, or something similar, to create a cobra-like appearance and thus make tourists open their wallets. The same thing has to be done if it actually is a cobra in the bottle.
The times when I have seen bottles with "cobras" in them for sale in Thailand or Cambodia, it has usually been the keelback. Buy a bottle of Singha instead and you will not contribute to the depletion of nature.
Pit vipers in alcohol do occur, but it is quite rare.

We should all climb down from our high horses and stop thinking that we Westerners are enlightened because we do not have such substandard animal entertainment in our home countries. It is true, but we must not forget that we have the choice not to. Our good economy gives us choices that are not available in all countries. Before you climb up on that high horse, you should ask yourself who is paying for the entertainment? Is it us or the Thai people who go to the "King Cobra show"? Spend your holiday money on other things, and those who work in tourism will work on more sensible things than treating snakes badly. Hire a wilderness guide! An adventure you will remember for a lifetime and a treat for your soul. Asia's nature is full of pit vipers and seeing an *Trimeresurus albolabris* in its natural habitat is amazing!

5. White-lipped pit viper

Trimeresurus albolabris (former *Cryptelytrops albolabris)*

In English, the *Trimeresurus albolabris* is called white-lipped pit viper, green tree pit viper, white-lipped tree viper, white-lipped green pit viper, white-lipped bamboo pit viper, bamboo pit viper and perhaps some names unknown to me.

In Thai, the species is called ngu khieow hạng mới thứng luệng (งูเขียวหางไหม้ท้องเหลือง). It is pronounced "ngu khieow hang mai thong lueang". You'll probably have to practice a bit to get that name pronounced right.

Conservation status

The *Trimeresurus albolabris* was last assessed for the IUCN (The International Union for Conservation of Nature) list of threatened species on September 1, 2011. The species was then considered to be viable and has since been listed as LC (least concern).

It is a term used when assessing the conservation status of species, i.e. when assessing how threatened they are. Species which are not judged to be facing major threats currently are placed into the category LC. These species are therefore not listed as red-listed.

Habitat

The species *Trimeresurus albolabris* has a large distribution area and you can find it in the following countries: Burma; Cambodia; southern China; Hong Kong; Indonesia (where isolated populations live in southern Sumatra and in western Java.); Laos; Malaysia (Peninsular Malaysia); Thailand and

Vietnam. It is not a species that migrates according to the season, as, for example, some populations of *Vipera berus* does in Sweden where I live.

This pit viper can adapt to a variety of habitats, and you can find the species both in dense forest and in big cities like Bangkok. They occur in many different environments, especially at lower altitudes (approximately 400 meters above sea level or lower) and the species thrives in both forest and scrubland, but it is usually assumed that the *albolabris* occurs up to 1200 meters above sea level. As well as in undisturbed habitat, you can find pit vipers in cultivated and urban environments such as on plantations, along roadsides, between open fields, near watercourses and in parks and gardens.

If you are looking for pit vipers in the wild, rainy season, when both predators and prey are more active, usually gives a slightly better outcome. Particularly common in low-lying areas with shrubbery and grassland and at the edges of cultivated fields. The species is rare in dense forests, and it is unlikely to find them in mangrove swamps and the similar.

Trimeresurus albolabris is a nocturnal species and avoids high daytime temperatures. You can still find this pit viper at any time of the day, but they usually become more active after dark. In addition, they are easier to spot with a flashlight in the dark because the green body color is reflected well by the glow of a flashlight. During the day, they blend in well with the green environment (see picture on next side).

While the species is arboreal, you often find them at a height of one meter or lower and you even find pit vipers on the ground. A good place to start looking is near waterways with some dense vegetation. As soon as it starts to get more humid, the chances of a successful "hunt" seem to increase.

Photo: Rickard Ljunggren. A female *Trimeresurus albolabris* photographed in the light of my headlamp. Photo taken with cell phone, it was raining that night, and the camera was not doing so well. The image is blurry but demonstrates well how the pit viper reflects the light of the lamp better than the leaves of the surrounding trees.

Photo: Jonathan Hagström (with my cell phone). A small *albolabris* we found one night on a rainy country road.

Description

Like other snakes, they are solitary animals, except during mating, and they live a life without social need or desire for closeness to others. I therefore choose to keep my pairs in separate terrariums, except during mating period. Also, it's easier to only have one venomous snake to keep track of when I open a terrarium.

The females become much larger than the males, they become both longer and, above all, much thicker. It is usually said that females grow up to 80cm and males up to 60cm, single females can approach 100cm, but that is unusual. The tail length for males is up to 12cm and for females 13 cm.

The *Trimeresurus albolabris* have a beautiful green color and my experience is that the males have a slightly brighter green color than the females. The side of the head below the eyes is yellow, white or light green and much lighter than the rest of the head.

Photo: Rickard Ljunggren. An adult female with a body that is much thicker than that of a male.

The belly is green, yellowish or white. The males have a thin white stripe along the side of the body. The tail of this pit viper is always reddish brown. Albinos are clearly yellow, but the reddish-brown tail retains its color. (see page 63).

Photo: Rickard Ljunggren. An adult male with his white stripe along the side of the body.

Photo: Rickard Ljunggren. An adult male,
with the slender body that males have.
Compare it with the body of the female on the previous page.

Photo: Rickard Ljunggren. An adult female (with an injured and dead eye). Few animals can look as grumpy as a pit viper can. In the picture you can also see the typical pits that pit vipers have.

Everybody in the reptile hobby probably knows that there are blue *Trimeresurus insularis*. It is probably a little less known fact that *Trimeresurus albolabris* also occurs in blueish color in the wild. However, it must be considered very unusual.

I have previously seen juveniles of this species that were partially blue, but the blue color faded with age and as adults the snakes were solid green. In the work with my book about the snakes in the *Viperidae* family (written in Swedish), I came across the phenomenon of single-coloured, blue *Trimeresurus albolabris*, where the blue color remains even in adulthood. It's a rare phenomenon, but for me, who has this species as something of a special interest, it was really interesting. The blue specimens of *albolabris* occur in only two known populations. One is in Siem Reap, Cambodia's second largest city which is known to many thanks to the ruin city of Angkor which is just outside the town. I've been there twice, but only encountered sleazy drug dealers, Burmese pythons and a lot of temple ruins.

The second population of blue *Trimeresurus albolabris* can be found on the Vietnamese island of Bạch Long Vĩ. This island is only three square kilometers in size and the nearest land area is just over seven kilometers away. Therefore, this *albolabris* population is probably fairly small and thus extra fragile if we start harvesting these snakes for the reptile hobby.

There are probably not any blue specimens of *albolabris* in the reptile hobby, at least not that reaches us mere mortals with wallets of normal size. I think that they will show up sooner or later, hopefully not by spoiling nature for its fantastically beautiful animals that are much needed in the food chain to help keep the balance of nature.

The *Crotalinae* subfamily, commonly known as pit vipers, or sometimes pit adders, are found in Asia and the Americas. Pit vipers are distinguished by the fact that their oval or triangular heads are provided with two deep pits, located between the nostrils and the eyes.

These pits are the external openings to a pair of extremely sensitive organs that detect infrared light. This gives the pit vipers what you might call a sixth sense that helps them see and find and perhaps even judge the size of any prey or threat. These pits work just as well at night as during daytime, they are completely independent of daylight. The pit organ has a complex structure and resembles the thermo-receptive pits found in boas and pythons. In boas and pythons, you can see these organs as a series of small dimples on the labial scales and in the case of *Aspidites*, behind the rostral scale. On page 50, the pits of a female *Trimeresurus albolabris* are clearly visible.

The underlying membrane is similar to a tympanic membrane that divides the pit into two sections of different sizes, with the larger of the two facing forward and exposed to the environment the snake is in. The two sections are inter-connected by a narrow channel that can dilate or contract by muscular contraction/relaxation.
Having one of these organs on each side of the head produces a stereo effect that indicates the position of the target. Experiments have shown that if pit vipers are deprived of their sense of sight and smell, these snakes can still strike accurately at moving objects that are less than 0.2 °C warmer than their surroundings.
Obviously, these organs are of great value to a predator that hunts at night and are also important in helping to avoid the snake's own predators. The snakes' heat-sensitive pits are also thought to help them find cooler areas in which to rest.

The *Trimeresurus albolabris*, just like other vipers, has rather long fangs in relation to its size. And just like other vipers, this species folds its fangs forwards when they need to be used. The fangs are used to assist the prey being swallowed. The pit viper articulates each fang forward alternately, pulling the prey down into the oesophagus. Once past the jaws, the prey is shifted by body movement, pushing the prey into the stomach.

Viper's fangs have the most efficient venom delivery among vertebrates and have been the focus of scientific interest for more than 200 years. Viper fangs have a venom duct that is formed by the fang sides closing against each other. together. Early development of the fangs of a juvenile *Trimeresurus albolabris* was studied. The study shows that the canal is formed by an early folding event, which is the result of an intussusception.

A pit viper's skin appears quite dull, just like the skin of the European viper, *Vipera berus*, is. And just like *Vipera berus*, the scales on the *Trimeresurus albolabris* head are many and small. The scales on the head are similar to each over, slightly overlapping, smooth or slightly keeled. The scales on the body are keeled as you can see on the picture below. Keeled scales have a ridge in the middle which is visible as a line on each scale. This ridge makes them rough to touch (see next page). Along the mouth opening on the upper jaw, the *Trimeresurus albolabris* has 10–11 supralabial scales. The term labial comes from *labium*, Latin for "lip". The supraocular scales that sit immediately above the eye are narrow, sometimes and enlarged. The size and shape of these scales are among the many characteristics used to distinguish species from each other. On the body, this pit viper usually has 21 rows of keeled midbody scales, slightly more or fewer may occur. These ventral scales are 155–166 in males and 152–176 in females. On the underside of the tail, the subcaudal scales are divided, 60–72 pairs in males and 49–66 pairs in females.

Photo: Rickard Ljunggren. Keeled scales on the body of a *Trimeresurus albolabris.*

In Southeast Asia, the red-tailed green rat snake (*Gonyosoma oxycephalum*) is sometimes confused with pit vipers, but just like the European *Natrix natrix*, the red-tailed green rat snake has a much shinier body than vipers do have. In terms of color, however, these snakes are similar as this rat snake also is green in body color, with a reddish-brown tail. *Gonyosoma oxycephalum* is of course much larger than a pit viper, but not all snakes are adult.

Photo: Photo Boards, Unsplash.

Threat

Contact with humans does not always end well for pit vipers. Since it is a venomous snake, it is not entirely unusual that humans feel the need to kill the animal they see as a potential threat. This is probably not enough to cause more than local population declines to this common viper. The species is unlikely to be affected by any other significant threats in its range as it readily adapts to modified habitats.

Photo: Rickard Ljunggren. A recently killed female *albolabris* in Huay Yang, Thailand. My left foot as a size reference.

If you drive around the countryside in Southeast Asia, you usually come across run-over snakes regularly. When I wrote this in the summer of 2023, we were in Huay Yang, which is just over an hour south of Hua Hin. I saw run over snakes almost every day when I was out driving.

Photo: Rickard Ljunggren. A run over juvenile monocle cobra, *Naja kaouthia*. This picture was also taken in Huay Yang, Thailand.

Pit vipers are captured, especially in Vietnam, and being used as food and traditional medicine (snake wine). Which species used is not always the same. However since *albolabris* is quite common in many places, it probably forms a significant part of this trade.
In many places in Southeast Asia, snakes are considered to have medical value, and often miracle potions and the like of it are considered to have a better effect if they are based on venomous snakes. For example, some believe that one can improve libido by drinking blood or eating a raw heart from a cobra. There are also spirits that are considered to improve libido, often with testicles from various animals as an ingredient. I myself, took a shot of Chinese liquor spiced with lizard testicles many years ago. Not because I expected any effect, but because I was curious about how bad it tasted, and it was really awful.
In Vietnam, some believe that snake wine can improve health and virility. Snake wine is drunk to cure everything from impotence to involuntary hair loss.

In southern Vietnam, the female pit viper is also captured because the embryos are eaten. Culture and tradition can sometimes lead people to eat odd things, especially if what you eat is attributed to some sort of health-promoting effect.

In Thailand, pit vipers appear in the pet trade, and it is not unusual to see *Trimeresurus albolabris* for sale in Europe with the locality "Thailand". I have also seen Vietnamese and Cambodian localities. The most commonly available variant of *Trimersurus albolabris* for sale is of Thai origin.

These examples may each represent a localised threat, but the species is probably not subject to overexploitation.

6. Care sheet

For many in the reptile hobby, the *Trimeresurus albolabris* is the first venomous snake they acquire. They are venomous snakes of a small size and they are not as quick or as toxic as many elapids. In addition, it is a species that is quite common in the hobby, the prices are reasonable and there are others who already have the species to consult if you need to. There are plenty of people in the arboreal viper hobby who specialize in them, and they will probably agree that the *Trimeresurus albolabris* is a good species of arboreal viper to start with if you want to keep venomous snakes. The reason they are so suitable as beginner vipers is that they are generally hardier than many other species and can withstand temperature and humidity changes better than most other arboreal species. I cherish the hobby and I think that some are in too much of a hurry to get venomous snakes, with hardly any experience keeping nonvenomous. Having venomous snakes is a hobby that is often questioned. Some think that it should be forbidden to have venomous snakes at home. Each time somebody gets bitten, or a snake escapes a private keeper, it can be the catalyst for a discussion regarding prohibition. Be responsible, do not rush and take care of your fingers!

Photo: Rickard Ljunggren. The door to the room where I keep my venomous snakes.

Trimeresurus albolabris is a species that requires a tropical environment in its terrarium. You need to use the entire height of the terrarium as it is an arboreal species that wants lots of branches to climb on and a higher place to lie down. Try to arrange so that there is a branch at a higher height both under the heat lamp and away from it.

Photo: Rickard Ljunggren. An adult male *albolabris* lying safely in the shade of the foliage.

As substrate, I use a thick layer of natural peat with a layer of leaves on top as this helps to keep the humidity up and works well to plant in. I buy the peat in plant shops in 40-litre bags. I think that the peat looks good in terrariums, and it is also cheaper than the substrate you can buy in zoo stores or at reptile fairs. I pick the leaves every autumn and store them in paper bags. I like to use leaves from wisteria for tropical terrariums as I like the look of these, and they last a long time before decaying. What you use is a matter of taste, pick what you can get your hands on.

Photo: Rickard Ljunggren. A male juvenile watching me when I come and bother him to take a picture.

Of course, you should use live plants and preferably have plenty of greenery! The plants create hiding places and security for the snake, it looks nice, and the plants tells you if it starts to dry out in the terrarium. I almost always use money plants as a complement to other plants in my terrariums. When the leaves no longer point upwards, it is time to pour a jug of water into the substrate and thus raise the humidity in the terrarium. Money plants, small-leaf spiderwort and monstera are examples of cheap and easy-care plants that cope well with the environment in a terrarium. For those of you who have a green thumb, there are plenty of tropical plants to buy at reptile fairs and sometimes in plant shops.

Photo: Rickard Ljunggren, unfortunately a somewhat blurry picture taken with my cell phone. A Malayan pit viper, *Calloselasma rhodostoma,* I found on a busy country road late one evening on the outskirts of the city of Hua Hin. The snake was captured and after a short ride on my motorcycle it was released in the forest outside the city the same evening. As you

can see, there is plenty of greenery in the forests of Asia, so don't scrimp with vegetation. No pit viper feels safe if you decorate as if you designed a golf course.

Photo: Rickard Ljunggren. A male *albolabris* hiding in the greenery of the terrarium.

Since they are, after all, venomous snakes, I recommend that you decorate the terrarium in such a way that you can always see the snake in its hiding place. The snake must feel secure, but you still always need to be able to see where it is located. For me, it feels good to always be able to see each snake in all the terrariums when I walk around in my venomous snake room. Even though I may not need to open the terrariums to do anything, it's always nice to have confirmation that the snake is where it should be and appears to be doing well. They usually say that you shouldn't bite the hand that feeds you, but that's a saying that hasn't reached pit vipers. If you are careless, sooner or later you will likely get bitten.

They are fairly defensive animals that can strike repeatedly when they feel threatened. They can also travel up a snake hook easily, so you must always be careful when handling.

Photo: Rickard Ljunggren.

Always label your terrariums with informative warning signs! In the event of a bite, it should be easy to tear off the information sign and take it with you when you seek care. Of course, you should always use a snake hook when handling and have a lock on a terrarium with venomous snakes.

Otherwise, this book does not contain any tips or advice on general venomous snake handling. If you want to learn this, reading a book is not the best way.

The knowledge and experience that one should have in order to reasonably safely handle potentially life-threatening snakes in the home cannot be acquired by reading my book.

Photo: Rickard Ljunggren.

It can be a good idea to have a mentor or attend venomous snake courses to learn what you need to know to be able to handle and care for a venomous snake in a safe and correct way.
Search for herpetological societies locally, where you will find serious people with great experience and a lot of knowledge about both nonvenomous and venomous snakes. Our Swedish society organizes field trips, social gatherings, lectures and joint trips to reptile fairs. The association also does important advocacy work and works for the good of our hobby when changes in laws and regulations risk affecting our hobby in a negative way. If you encounter problems with your snake, there are always society members who can help you. No matter what your problem is, you are guaranteed not to be the first to have it. My advice is to find a herpetological society and join it.

Sveriges Herpetologiska Riksförenings logo, a *Bitis nasicornis.*
(Our Swedish herpetological society.)

Photo: Dennis Sjöberg. An albino *Trimeresurus albolabris*, a female. An albino male would still have his white stripe along the side of the body. Note that the red-brown color of the tail is unchanged. Albinism also occurs in nature, but there the snakes are more vulnerable when the camouflage is out of order.

Photo: Rickard Ljunggren. Mating *Trimeresurus albolabris*. Except during mating, my pit vipers always have separate residences. Males can be a bit difficult feeding if they share accommodation with a female. (The female in the picture has an eye that is dead from an old injury.)

The terrarium

How big of a home do you need for your *Trimeresurus albolabris?* In Sweden, we have statutory minimum requirements based on the snake's size and lifestyle. In other countries, one might think that the accommodation should at least be as long and high as the length of the snake and have half the length of the snake as depth? Think "the bigger the better", that is always a good motto when buying snake enclosures! A well-proportioned and well-decorated terrarium is an ornament and creates opportunities for the snake to seek the microclimate that is optimal for the moment. Make sure to decorate so that you use the entire height of the terrarium and also make sure to fill the terrarium with decor. No snake wants to live in an empty apartment with a lot of unnecessary ceiling height!

Photo: Rickard Ljunggren. Terrarium for one of my female *Trimeresurus albolabris*. Since the picture was taken, the newly planted money plant has taken off and it has almost become a small jungle in the jar. The small-leaf spiderwort grows like a weed and needs to be cut quite often! The ground is covered with several hiding places, in case she chooses to spend her time there.

Humidity

The *Trimeresurus albolabris* lives in Southeast Asia and requires a tropical climate in the terrarium. The humidity should be 65–80%. If you often need to spray water in the terrarium to keep the humidity up, my suggestion is that you go over the snake's accommodation. As I wrote in the beginning of this chapter, I have a thick layer of bottom substrate, live plants and lots of decor as this helps keep the humidity up. I often have ten centimeter of peat at the bottom which is constantly kept moist. Ventilation is good, but if you have too much ventilation, the environment in the terrarium will soon dry out. If you instead reduce the ventilation too much, the humid climate can create problems with mould in the terrarium. Do not spray water in the terrarium too often, the snake does not need to live in a constantly recurring drizzle.

You can, however, spray a little extra or make sure that the bottom substrate maintains a higher humidity to facilitate the snake's shedding. In addition, it's nice for you not to have to peel off the remaining shed of a pit viper that doesn't appreciate your company.

A large water bowl helps with the humidity and gives the snake the opportunity to bathe if it wants to. It is, however, something that I have never observed any pit viper do.

Heat

27–29 °C during the day may be considered an optimal temperature. Of course, you should have the heat lamp positioned so that there is both a warmer and a cooler zone in the terrarium. If your arboreal pit viper is always on the ground in the terrarium, you probably need to lower the heat and the easiest way to do this is to dim the heat lamp or use a weaker lamp.

When the lamp is switched off in the evening, it becomes a natural nighttime lowering of the temperature. However, ensure temperature so that it doesn't get too cold before morning, 25–26 °C is ideal at night. If you need to, you can use a heating mat as supplementary heat, although it can dehumidify the lower substrate. For me, the night lowering is easy as I have a venomous snake room with a higher temperature than what we have in the rest of the house.

Diet

This pit viper's wild diet includes small rodents, lizards, birds and frogs. In search of prey, *Trimeresurus albolabris* often rests on a branch with its focus on the ground below, waiting for a prey animal to be unlucky enough to pass by too closely. If you see your pit viper positioning in a similar manner, c feeding is usually successful. If you have a male and a female in the same terrarium, it may happen that the male fusses with the food or stops eating altogether. If your pit viper is too small to swallow a whole pinkie, you can give it a leg from an adult mouse or a head from a pinkie.

Photo previous page: Rickard Ljunggren. Picture taken in a fairly newly planted terrarium, so the vegetation has not really started to grow yet. Here you can see a hungry male hanging on a branch waiting for prey to pass by on the ground below.

If they come across prey, a hungry *Trimeresurus albolabris* often will strike with lightning speed. Once prey is captured, the pit viper will hold on to it unlike some terrestrial venomous snakes who release their prey and then wait for it to succumb to the venom before recapturing it. Other arboreal venomous snakes also have this behavior when it comes to holding on to the prey so that it does not disappear somewhere on the ground when the snake has released its grip. Terrestrial species who prey on birds, reptiles and amphibians also tend to hold on to their prey.
A trick I often use when feeding pit vipers is to thaw the food in hot water. Take the hottest you can get from the tap (about 60 °C is a normal maximum tap temperature here in Sweden), but not water you heat up. The food should still be raw. Allow the mouse to cool down for a few seconds before offering the snake food.
Feed the snake at a time of day that matches the snake's natural circadian rhythm. I always feed my pit vipers in the evening but nighttime works just as well. Hold the prey in front of the snake's face or stroke the prey lightly along the snake's mouth. You can also try touching the snake's tail with the prey as this can sometimes make the snake bite.
One of my pit vipers prefers that I put a mouse on a branch in the evening and the next morning the mouse will be gone.
(I feed my snakes with dead prey, always.)
Males can be a bit more difficult to succeed with. It can sometimes take several defensive bites before the male viper decides to hold onto the prey. Juveniles of both sexes can sometimes react similarly if they see the tweezers or the mouse as a threat instead of seeing the mouse as food and ignoring the tweezers.

Getting a nervous juvenile to eat can be a test of your patience. Sometimes it can help if the snake doesn't see you. Sometimes I hide behind a transparent lid from a bigger plastic tray or make sure to sit still. At worst, it can take half an hour to get a juvenile pit viper to eat the first few times, but you can see it as an unusual form of meditation.

You will soon find out what suits your pit viper best.

Photo: Rickard Ljunggren. A female that has taken her prey. The female in the picture is older than you might think. She is a dwarf and when the picture was taken, she was three years old and still eating pinkies. The food works perfectly, and she eats every time I try to feed her, but she is almost not growing at all. My guess is that she won't get any bigger. An odd female viper, but for her sake I won't try to get her to breed.

Water

It is not entirely unusual for a pit viper to not want to drink from a water bowl and especially newborn pit vipers may ignore the bowl of water in the terrarium. They then prefer to drink water drops from their body, from branches and leaves in the interior or from the glass panes of the terrarium. If you have pit vipers that do not drink from water bowls, you should spray the snake and interior every day. But do not spray too much so that after a few days of spraying it becomes soaked in the snake's habitat. Even if your snake does not want to drink from a water bowl, you should still have one with fresh water in the terrarium. It helps to keep the humidity up and it could happen that your pit viper discovers the benefit of drinking from a bowl.

Mating

Males are ready to mate at a fairly small size, while females will only consent to mate when they are of sufficient size. This may take two years or more. Mating usually takes place from August to October and during this period the male may stop eating completely for several months. The male will not eat again until he has moved back to his own terrarium. Always let the male move in with the female. Never do the opposite, it is the male who should seek a female to mate with. If your male is a bit slow starting, it can help to put a shed skin from another male into the female's terrarium. The skin can trick the male into thinking there is competition and make him take the opportunity to mate while the opportunity exists.

A litter usually contains 10–15 live young (not eggs), but exceptional litters of 25–30 young have occurred. It usually takes 129–157 days from a successful mating until the female give birth. You can see a picture of mating Trimeresurus albolabris on page 64.

Photo: Rickard Ljunggren. Unfertilized egg from a *Trimeresurus albolabris.*

Age/Longevity

In the wild, a *Trimeresurus albolabris* can live approximately until 10 years of age, in a good environment in a terrarium it can live much longer than that. It is important to housing the species correctly in the right conditions that include a night drop. Also, be sure not to overfeed your pit viper. Fat and overfed snakes are common in the hobby, and it significantly reduces the expected lifespan. Females in particular can be quite happy to eat and are easily over fed.

7. Venom

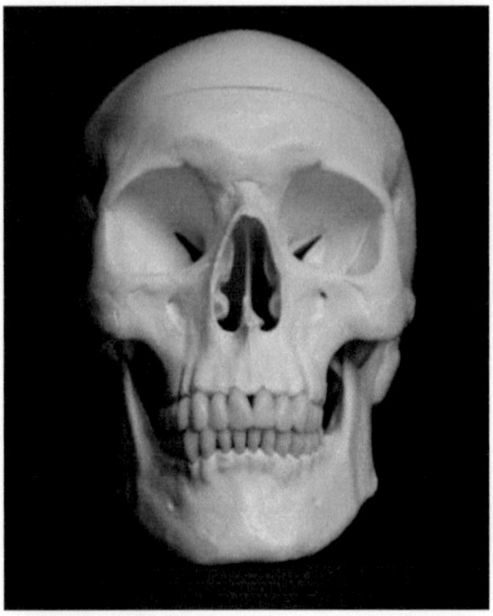
Photo: Anne Nygård

Pit vipers usually prefer to strike rather than flee and are quick to bite if you happen to get too close to them or if you are touching them. This behavior is seen in many species of vipers. They are not so mobile and quick to escape and because of that they often choose to lie still with the hope of remaining undetected. If this fails, an attempt to bite may be the only option left for the snake.

The average amount of venom per bite from a *Trimeresurus albolabris* is 8–15 mg (dry weight).
The strength of the venom according to LD50 is 12.75 mg/kg subcutaneously.

LD50 numbers are often used as a general indicator of a substance's acute toxicity. A lower LD50 indicates increased toxicity. As a comparison to the *Trimeresurus albolabris*, arsenic is 14 mg/kg subcutaneously and the venom of the inland taipan (*Oxyuranus microlepidotus*) is 0.025 mg/kg subcutaneously.

The venom of *Trimeresurus albolabris* is primarily haemotoxic. Haemotoxins are used by a variety of animal species, including pit vipers. In addition to killing the prey, part of the function of a haemotoxic venom for some animals is to aid digestion. The venom breaks down protein in the area around the bite, making the prey easier to digest.

Haemotoxins are toxins that destroy red blood cells, interfere with blood coagulation and/or cause organ degeneration and tissue damage around the bite. Injury from a haemotoxic agent is often very painful and can cause permanent damage and, in severe cases, death. General envenomation rate is unknown but is likely to be high according to those who know how to assess such things. The process by which a haemotoxic venom causes death is much slower than that of a neurotoxic venom such as the elapids (cobras, mambas, etc.) have.

Snakes that envenomate a prey may need to track down the prey as it flees. Usually, prey will soon stop escaping due to shock caused by the venom. Although the prey is still alive, it is no longer able to continue escape. Arboreal vipers usually do not need to search for their fleeing prey as they rarely release the prey after delivering a bite.

A bite from *Trimeresurus albolabris* can mean anything from mild envenomation to death in rare cases. Bites from pit vipers are not entirely uncommon in Southeast Asia but there are very few deaths reported. Loss or limited function of an affected limb is possible even with prompt treatment. Milder symptoms after a bite can be nausea, disorientation and headache. These symptoms may be delayed for several hours.

Possible effects of a bite, in detail:

*Locally around the bite, you can expect severe pain, swelling due to accumulation of fluid, bruises, blisters and possibly necrosis (tissue damage) in rarer cases.
*The body's systems may react with nausea, vomiting, diarrhea, headache, dizziness, abdominal pain, racing heart and racing pulse, abnormally low blood pressure, collapse and convulsions.
*Heavy bleeding and increased coagulation.
*The body can go into shock due to fluid loss and tissue damage.
*An elevated white blood cell count, which can be caused by stress and infection.
*Hypokalemia, a condition of abnormally low levels of potassium ions in the blood. It is a serious condition that requires immediate treatment, as many processes in the body depend on stable levels of potassium ions in the blood.
*Hypocalcemia, low calcium levels in the blood. Hypocalcemia causes tingling and pins and needles in the body as well as tetany, a condition prone to convulsions. The central nervous system is affected and hypocalcemic individuals are often easily agitated and confused.
*Myoglobin and blood in the urine. Myoglobin is the most important oxygen-binding molecule in muscle tissue.
*Possible secondary infections.

Other effects may possibly occur.
Dry bites may occur and then, of course, you avoid all of the above.

8. If you get bit

Perform the following procedure without any delay:

1. Ensure that the snake is secured, either in its terrarium or in other closed, locked and secure storage.

2. Wash the area around the bite to remove any venom on your skin.

3. Keep calm and call the emergency number. Provide emergency care with the full name of the species to ensure appropriate treatment.

4. Remove any tight clothing or jewelry pending any swelling.

5. Get to the nearest emergency room, preferably via ambulance or immediate transport. Don't drive yourself! If you have a sign on your terrarium with the species name, or something similar, take this with you.

6. Don't use a tourniquet or apply ice. Don't cut the bite or try to remove the venom. You are not a doctor!

7. If possible, try to keep the bitten body part below the level of your heart so that the venom does not spread with the blood circulation more than necessary.

8. Mark the leading edge of tenderness/swelling on the skin and write the time alongside it.

9. Thanks!

Thank you for choosing to read my book! I hope you find the *Trimeresurus albolabris* as beautiful and fascinating as I do!

Photo: Rickard Ljunggren. A male *Trimeresurus albolabris.* If you look closely, you can see the pink tongue of the species.

Enjoy nature in Asia if you have the opportunity to do so but do it with respect for both nature and your own health.

Photo: Jonathan Hagström. Me watching a *Naja siamensis* in the nature surrounding Hua Hin, Thailand.

It is often a good idea to be two or more in the field, in case something happens. It doesn't have to be a life-threatening snakebite, a sprained foot can be just as problematic if you are alone far out in the forest. If you don't feel completely confident going out on your own or with friends, there is almost always access to local nature guides.
Take pictures, leave footprints.

Pit vipers are fantastically nice to keep in a terrarium. They are often fully visible up on a branch focusing on the bottom of the terrarium, in case a prey comes along. They are also snakes that do not become terribly large and long. They are snakes that are reasonable to work with that also do not reach too far with their strikes. As I write these lines, I have specimens of the Asian pit viper species *albolabris, insularis, macrops* and *venustus* in my terrariums. All species are equally beautiful, and it is difficult to choose a favorite!

So, who am I?

I am almost 50 years old and I live in Hässleholm, in the south of Sweden. Sweden is a country that only has three snake species, of which *Vipera berus* is the only venomous one.
I have been interested in snakes since I was a child, the *Vipera berus* was the snake species that got my attention. Where I grew up, we were close to the forest and I previously captured *Natrix natrix* and *Anguis fragilis* (not a snake, a lizard without legs) and took them home and kept them in the garden for a while before it had to move out into the forest again. Of course, you must not catch wild animals to take home. I know that today, but that knowledge was lacking when I was growing up.

The third species of snake we have in Sweden is *Coronella austriaca*, but they can't be found near Hässleholm.

Getting a snake as a pet wasn't an option as long as I lived with my parents, so my first reptile was a red-eared slider (*Trachemys scripta elegans).*
When I got my own apartment, I got my first snake and since then I've probably owned about 200 snakes. My first snake was a common garter snake (*Thamnophis sirtalis*) and I still love garter snakes.

Today I have about 35 snakes in my basement, both harmless and venomous species.

Photo: Jenny Ljunggren. Me down in our basement with a large female Trimeresurus albolabris on my snake hooks.

I enjoy photographing wild vipers here in Sweden. I also travel to Southeast Asia as often as I can afford to. Nothing beats herping in Thailand with a headlamp, looking for snakes at night! It is an exciting adventure!
As I write these closing lines, I have my next trip to Thailand two weeks ahead of me. Two of my goals with this trip are to see *Trimeresurus macrops* in the wild and of course to see *Trimeresurus albolabris* again.

This is my eighth book about snakes (one of the earlier being a Swedish version of this book) and my first attempt at writing in English. If it turns out well, I will release more books in English. The subject is interesting, and I am learning more as I write. Hopefully I will improve my English writing gradually.

Photo: Rickard Ljunggren. A melanistic male *Vipera berus*, the snake species that got me interested in snakes. Picture taken outside Hässleholm, Sweden.

Thank you once again for choosing to read my book! I hope you get to experience an encounter with these exciting animals in the wild or get the opportunity to keep *albolabris* in a terrarium!

Best regards, Rickard

Licensing [edit]

This work is in the **public domain** in its country of origin and other countries and areas where the copyright term is the author's **life plus 100 years or fewer**.

⚠ You must also include a United States public domain tag to indicate why this work is in the public domain in the United States.

This file has been identified as being free of known restrictions under copyright law, including all related and neighboring rights.

Page 15

Licensing [edit]

I, the copyright holder of this work, hereby publish it under the following license:

This file is licensed under the Creative Commons Attribution-Share Alike 2.5 Generic license.

You are free:

- **to share** – to copy, distribute and transmit the work
- **to remix** – to adapt the work

Under the following conditions:

- **attribution** – You must give appropriate credit, provide a link to the license, and indicate if changes were made. You may do so in any reasonable manner, but not in any way that suggests the licensor endorses you or your use.
- **share alike** – If you remix, transform, or build upon the material, you must distribute your contributions under the same or compatible license as the original.

Page 29

Licensing [edit]

This file is a work of a U.S. Army soldier or employee, taken or made as part of that person's official duties. As a work of the U.S. federal government, it is in the ***public domain*** *in the United States.*

العربية · বাংলা · català · čeština · Deutsch · English · español · eesti · فارسی · suomi · français · hrvatski · magyar · Bahasa Indonesia · italiano · 日本語 · 한국어 · lietuvių · македонски · മലയാളം · မြန်မာဘာသာ · Nederlands · polski · português · русский · sicilianu · српски / srpski · Türkçe · українська · Tiếng Việt · 中文（简体） · 中文（繁體） · +/-

Page 33

Licensing [edit]

This work is in the **public domain** in the United States because it is a work prepared by an officer or employee of the United States Government as part of that person's official duties under the terms of *Title 17, Chapter 1, Section 105 of the US Code*. **<u>Note</u>: This only applies to original works of the Federal Government and not to the work of any individual U.S. state, territory, commonwealth, county, municipality, or any other subdivision. This template also does not apply to postage stamp designs published by the United States Postal Service since 1978. (See § 313.6(C)(1) of Compendium of U.S. Copyright Office Practices). It also does not apply to certain US coins; see The US Mint Terms of Use.**

This file has been identified as being free of known restrictions under copyright law, including all related and neighboring rights.

Page 34